J. Schroeter, G. Best

## New Observations in Further Proof

of the Mountainous Inequalities, Rotation, Atmosphere, and Twilight, of

the Planet Venus. By John Jerome Schroeter, Esq. Communicated by

George Best, Esq. F. R. S.

J. Schroeter, G. Best

**New Observations in Further Proof**
of the Mountainous Inequalities, Rotation, Atmosphere, and Twilight, of the Planet Venus. By John Jerome Schroeter, Esq. Communicated by George Best, Esq. F. R. S.

ISBN/EAN: 9783337382131

Printed in Europe, USA, Canada, Australia, Japan

Cover: Foto ©berggeist007 / pixelio.de

More available books at **www.hansebooks.com**

V. *New Observations in further Proof of the mountainous Inequalities, Rotation, Atmosphere, and Twilight, of the Planet Venus.* By John Jerome Schroeter, *Esq.* Communicated by George Best, *Esq. F. R. S.*

*(Translated from the* German.)

Read February 19, 1795.

## PREFACE.

ALTHOUGH it is a satisfaction to me, that Dr. HERSCHEL last year found my discovery of the morning and evening twilight of Venus's atmosphere to be confirmed, as I could not hope to have obtained such an important confirmation so early, considering the excellent telescopes required, and that a favourable opportunity for such observations occurs but rarely ; yet the paper *on the Planet Venus,* which this great observer has inserted in the Philosophical Transactions for 1793, contains unreserved assertions, which may be easily injurious to the truth, for the very reason that they have truth for their object, and yet rest on no sufficient foundation.

Openness, without reserve or indirect views, must guide the spirit of observation in the true inquirer into nature, and be his sole object. To this pure source alone can I ascribe what is said in the abovementioned paper, so as to reconcile it to the friendly sentiments which the author has always hitherto expressed toward me, and which I hold extremely precious ;

though perhaps to others it may not have the same appearance. But this very object makes it also my duty to be equally unreserved in remarking what truth is, and demands; particularly as evident misunderstanding and error appear to have chiefly occasioned those assertions; which most probably would not have been thus made, if the author had then known of my very circumstantial memoir,* which was read at the jubilee of the university of Erfurt, in a meeting of the Electoral Academy of Sciences, and which they ordered to be printed; and could have compared the many careful observations, full of matter, contained in it. A copy of this memoir I have lately had the honour of communicating to the worthy author of the abovementioned paper.

Therefore, in order to prevent misapprehensions, let me be allowed to make some remarks, which truth requires of me, before I communicate faithfully, as I mean to do, my more recent observations, which confirm the former ones, and seem to me very important.

1. The celebrated author considers it, *with reason*, as a wonderful relation, that I should profess to have seen *appearances of spherical spots* on Saturn, without having, at the same time, determined from them the period of his rotation, which might have been done in the first hour; and he thinks that no one, who is not possessed of incomparably better sight and telescopes than he has, can have seen any thing of the kind. In that I fully agree with him, and here declare publicly, *that I have never perceived such an appearance on Saturn, however much I wished it.*

---

* *Beobachtungen über die sehr beträchtlichen Gebirge und Rotation der Venus,* with three copperplates. Erfurt, 1793.

In the German original of my paper, the translation of which is published in the Philosophical Transactions,* it stands thus : " On the contrary, from the circumstance that NO SUCH " EVIDENT FLATTENED SPHERICAL FORM *is perceived in this* " *planet* (namely, at its poles) *as in Jupiter and Saturn*," &c. &c.

The author indisputably agrees with me in all the truths there asserted. He has himself observed the flattened shape of Saturn at the poles more exactly than I, and even determined the proportion of the shorter to the longer axis. But in the translation, for the words " *abgeplattete kugelgestalt des Jupiter* " *und Saturn*," is put " *flat spherical forms*," &c. which he understood as if I pretended to have observed spherical spots on Saturn. The author might have convinced himself of the contrary, by comparing the German original in the possession of the Royal Society.

2. He considers it as an equally wonderful relation, that I have SEEN in Venus, in the same manner as in the moon, mountains and shadows of mountains, which were four or five times higher than our Chimboraço, and that I thence pretended to have determined the rotation of this planet ; on the contrary, he considers this last as hitherto undetermined, *because* HE *has never found a trace of mountains*, and all his observations, for 16 years past, have been absolutely insufficient to ascertain it, *though nothing of that kind could well have remained hid from him.*

Here it is not myself, but the truth, that I undertake to defend ; and I am convinced that if my memoir above mentioned, on the Rotation of Venus, had been already known,

* Observations on the Atmospheres of Venus and the Moon; their respective Densities, perpendicular Heights, and the Twilight occasioned by them. Phil. Trans. 1792.

and the author had compared the almost innumerable and various observations contained in it, which all agree in their result, he would never have made such a declaration. *I have myself also never actually* SEEN MOUNTAINS *in Venus* AS IN THE MOON; but only *deduced* their existence and height from the observed appearances. It is even impossible to see them, according to what I have expressly asserted in my paper on the Twilight of Venus ; because, on account of the thickness of her atmosphere, we can never perceive the shades of land on her surface. But if the appearances observed by me and others are true, the result deduced from them is mathematically evident.

That I have seen, *not unfrequently*, the boundary of illumination irregular, is *nothing new*, nor does it afford me any further merit than that of *confirming* with many others, *an old truth*, which DE LA HIRE, and still more ancient good astronomers, provided with the best and most powerful telescopes of their kind, had long ago discovered in perfectly similar phænomena. So early as the year 1700, DE LA HIRE observed greater inequalities in the termination of light in Venus, than in the moon ;* and the Paris Academy thence concluded that planet to have higher mountains. The sole addition, as far as I know, which I have made to the older observations is, that in the crescent phase of Venus, sometimes one horn is only half as broad as the other ; and that sometimes, though not often, about the period of the greatest elongation, one end of the enlightened part appears pointed, but the other rounded off: appearances which others, who had not been apprized of what they were to see, have frequently perceived as well, and

* See *Mémoires de l'Acad. des Scienc.* 1700, p. 378.

in the same manner as myself. It is here scarcely necessary to remind the reader, with respect to the ancient observations, that in all those where no extraordinary light is wanted, particularly powerful telescopes are by no means required. I should indeed be surprised that the celebrated author had not, in all the time since 1777, perceived any inequality in the boundary of light, or other appearance of that kind, tending to confirm the existence of very high mountains according to the old observations, were it not that his bold spirit of investigation has been chiefly employed in making much more extensive discoveries in the far distant regions of the heavens, where he has gathered unfading laurels. In fact, the observations which he has communicated from his journal are *much too few* to prove a negative against old and recent astronomers. Without encroaching upon truth in the least, I could certainly produce more good distinct observations during many months, from 1779, when I began to examine Venus carefully, to 1793, when my memoir on her rotation was finished, than are adduced for a period of 16 years in the abovementioned paper of my opponent: having, in the latter years, observed this planet not only daily, but, as far as the weather and her position admitted, almost hourly through the whole day and evening. This, I think, is shewn evidently enough by the memoir already mentioned, in which only the later observations appertaining to the subject are inserted : and without such steady perseverance, my trouble for so many years would have been fruitless, as was the case with other observers ; *for, in almost innumerable observations, the same thing happened to me as to the author of the paper in question, namely, I perceived neither spots, nor any other remarkable appearance, except the unusually quick*

*decrease of light toward the boundary of illumination, which itself
was not sharply defined.*

It is right that every acute observer should be on his
guard against a precipitation which often occurs, and not con-
tradict respectable astronomers who have preceded him, if
he should not at once, in a few observations, find those appearances in an object which such credible men have perceived, or deduced from their observations. The mischief
thence arising may be important, and lead to more general
error in proportion to the celebrity of the contradicting observer, because there are always persons enow who will adopt
it as a truth without further examination. And yet there are
many examples of this in the most modern history of astro-
nomy. Thus, for instance, the old worthy selenographer
HEVELIUS found some of the mountains of the moon to be
more than $\frac{2}{3}$ of a (German) geographical mile in perpendicular
height; and this truth stood more than 100 years in all the
elementary books. Later astronomers measured only a few
of those mountains, and partly not with all the requisite cir-
cumspection ; yet concluded, from too few and insufficient
observations, that HEVELIUS had given them much too high.*
This was already received as true in the elementary books ;
notwithstanding which the excellent HEVELIUS was absolutely
in the right, as is proved by my numerous and incontrovertible
measurements.†

When, in the years 1789 and 1790, the ring of Saturn appeared as a straight line of light, I perceived only a few pro-

---

* See RÖSLERS *Handbuch der practischen Astronomie,* 1 Th. p. 441.—Philos.
Trans. Vol. LXX.

† *Selenographische Fragmente,* § 34 to 82.

jecting luminous points on it till after October, 1789; but in February, 1790, incomparably more of them, in the frequent observations I made. These, in part at least, I considered not as satellites, but as true and large inequalities of the surface of the ring; and thence drew, on the strongest grounds of probability, the same conclusions as MESSIER and other respectable observers had done 15 and 30 years before; one of those deductions, and which seemed highly probable, was, *that the southern surface of the ring must have many more and larger inequalities than the northern.* These remarks had already been made known to the world, in the publications of the Naturalist Friends at Berlin; * when I unexpectedly read in the Philosophical Transactions a conclusion which discouraged me very much, that the astronomers who considered these projecting luminous points as inequalities of the surface, were mistaken, those appearances being occasioned by the satellites of Saturn: this conclusion was drawn from some new and excellent observations inserted in the paper itself, but which were continued only to November, 1789. However, so much the greater was my pleasure to find this assertion recalled in the next volume of the Transactions, where Dr. HERSCHEL, from those very projections, has made the important discovery of the rotation of the ring, and determined its period. Now, if there are really in the ring of Saturn such enormous inequalities, I do not see why my conclusion, deduced from so many agreeing observations, namely, *that the mountains of Venus bear nearly the same proportion in height to her diameter, as those of the moon do to the diameter of the moon,* should be thought a wonderful relation, especially since all my

* *Schriften der Naturforschenden Freunde.*

observations hitherto, as for instance those on the visible luminous spots in the dark part of the moon, on the apparent
changes of the moon's surface, &c. have been confirmed by
others.

From these remarks, the answer will readily present itself,

3. How the author of that paper could look upon my observations on the rotation of Venus as unfounded, though
there are so many of them which agree together, and *he had
not read and compared them;* and could think the period of rotation as much undetermined as before. Whoever deigns to
bestow some attention on my memoir on the rotation of Venus, will soon find,

(*a*) That certainly I did not go to work carelessly, *but first*
arrived *gradually* at an *approximate* estimation *by almost innumerable observations made in very different ways.*

Although I perceived, as early as in the year 1786, some luminous spots of Venus, which seemed to me to shew a period of
rotation of about 24 hours, as Dom. CASSINI had also thought ;
yet I suffered them to lie unpublished six years, because I was
doubtful whether some delusion might not have intermixed
itself; until at length a favourable opportunity accidentally
led me to pursue the investigation of this subject *in an entirely
different manner.*

(*b*) It will also be found that the author, among his observations, which taken altogether are *but few, cannot shew a
single one in which he observed at the same time with me.* But
every person conversant in these subjects will agree with me,
that in order to prove the inaccuracy of my observations, or at
least render them doubtful, it is essentially necessary, *that an
impartial observer should have* DIRECTED HIS ATTENTION WITH

EQUAL CARE TO THE SAME CIRCUMSTANCES AT THE SAME TIME, *and not have seen them the same as I have given them.* In my memoir, to which I here refer, those observations only which belong to the point in view are compared together ; *but in other observations, almost innumerable, which I made partly before I had paid any particular regard to the inequality of the horns, and partly in the intervals, I did not perceive, any more than the author, either spots or any thing appertaining to the matter in question ; and consequently our corresponding observations perfectly agree together.* It is, however, and will remain a truth, that there is no such thing as a monopoly of discoveries ; one man may luckily observe something to which the other did not direct his attention *in the same manner,* although he viewed it at the very same moment. Thus, for instance, since HEVELIUS's time many observers, provided with sufficiently powerful telescopes, have examined the moon, without perceiving the immense southern *cordilleras* of her edge, the perpendicular height of which, by indisputable observations, amounts to something more than a geographical mile, and which I have pointed out and delineated in my Selenotopographical Fragments, under the names of Leibnitz and Doerfel. And yet these high mountains are really there, and afforded a magnificent spectacle at the commencement of the solar eclipse on the 5th of September last year, though they were not then exhibited in their greatest projection. So likewise it is true, that several of the many important discoveries, on which the author has founded his eternal fame, might have been made as well by other observers, who were furnished with good achromatic telescopes, if they had directed their attention in the same manner to the same objects, with equal acuteness and perseverance.

Having premised these remarks, I can now communicate exactly, and according to their connection, my new Observations on the Planet Venus; and that they may, in various points, be more easily and better compared with the observations of my opponent, I will at present follow the order of my journal.

*New Observations, confirming the Rotation of Venus, her mountainous Inequalities, and the Twilight of her Atmosphere.*

*Feb.* 18, 1793, $5^b$ 50' *p. m.* As cloudy weather had continued uncommonly long, and as the experience of many years had already shewn that little or nothing remarkable is to be expected, when considerably more than half of Venus is illuminated, I could not till this time proceed on the observations, the planet now approaching her greatest eastern elongation. With 160 of the 7-feet SCHRADERIAN telescope, I had, with the full aperture, such an extraordinary soft and clear image as I scarcely ever found in this planet. According to fig. 1. (Tab. XII.) both ends of the boundary of light appeared equally rounded, without any perceptible difference. There was, however, again, in the middle of the enlightened part, a kind of darker nebulosity, not quite clearly to be distinguished, which seemed to consist of two very slight nebulous spots. The light decreased to extraordinary dimness toward the boundary of illumination.

. *Feb.* 26, $5^b$ 15' *p. m.* An extremely remarkable observation. With 160, 288, and 370 magnifying power of the 7-feet SCHR.

I found, the image being uncommonly fine and soft, that as usual there was no spot, but that the northern end of the boundary of light, *a*, fig. 2. was most certainly rounded off beyond all comparison more than the southern ; the latter appearing to run on rather pointed, with an inequality upon it, on which a dim greyish shadow was perceived.

At 6$^b$ 20'. In order to secure myself against deception, I desired my attendant, who came in at that time, and has remarkably good sight, with some practice, to observe whether he saw any thing particular ; and what? The answer he gave, at the first sight, was, that Venus had *an evidently irregular form ; that on the right* (southern) *end of the illumination she was pointed, the point having some shade on it, but that on the left she was oval.*

At 6$^b$ 40', the difference began to be less striking ; and having intermitted the observation in order to recruit my eye, I found at 7$^h$ 30' both horns equally rounded, though with this difference, that at the southern one a small indistinct glimmering point of light, barely perceptible, often shewed itself at *a*, fig. 3. not on the rounded part, but close to it : this was seen with 288, as well as 160. At 7$^h$ 45' I found it still the same ; and likewise afterwards with the 13-feet reflector, which also shewed me the point. Soon after, Venus became invisible.

There was no nebulosity to be perceived as on the 18th.

*Feb.* 27. I wished much to examine the changes which might happen in the course of all this afternoon, but high light clouds prevented me. It was very remarkable, *that at 5$^h$ 40' on this succeeding day, I saw most distinctly the same appearance as the evening before,* with 109 and 160 magnifying powers, only with this slight difference, *that the shadow, which shewed itself* *again at the southern point, as at a, fig.* 4, *entered westward a little further into the point;* and it sometimes appeared as if the shadow would penetrate all the way through, and entirely cut off the point ; moreover the northern horn was not quite so much rounded as the evening before. With both magnifying powers I saw likewise again, on the illuminated part, a very faint oblong nebulosity *b*, distant only about $\frac{1}{3}$ of the semidiameter from the external edge. For greater certainty I applied a power of 288 and 370, with which I distinguished the abovementioned form of the illuminated part, extraordinarily fine and distinct ; I could likewise see, with all the magnifying powers, the darker indentation of shadow *a*, but not the very slight nebulosity *b*. The indentation of shadow was in length at least $\frac{1}{10}$ of the semidiameter ; and at 6$^h$ 11′ it began to pass quite through, so that the southern horn appeared rounded like the northern, and the fine point, being now separated, looked like a glimmering dot of light close to it. I saw this separate point of light repeatedly, with 209 times, among other magnifying powers, very plain and evident, the image being soft ; in different observations I found it always the same, whatever was the power ; and at 6$^h$ 19′ the southern end appeared fully as round as the northern. I thought it remarkable, that at 6$^h$ 25′, a power of 288 shewed it smaller than it appeared with a less power. At 7$^h$ 12′ the point of light had vanished, as I perceived with both the 7 and 13-feet SCHRADERIAN reflectors, Mr. TISCHBEIN, the instrument-maker, who came in toward the end of the observation, saw it in the same manner. Both horns at this time appeared quite equally rounded ; but a new remarkable circumstance was now first discovered by Mr. TISCHBEIN. He observed with both reflectors, that at the northern horn, though rounded like the southern, *a brighter pointed small inequality* projected out from the faint boundary of light, as is expressed at *a*, fig. 5. It was difficult to distinguish, but his eye, more accustomed to microscopic objects, saw it alike with both reflectors, and *in the same place*; I perceived it also, though it was not striking. The observation was continued by both of us to 8ʰ 30′, when Venus being sunk too low, began to be indistinct. At this time indeed I could no longer distinguish that fine point; but in every part of the field of the instrument something brighter appeared in its *fixed* place.

Whoever is pleased to compare these two observations impartially, I doubt will not consider them as illusions. To me they rather appear, in more than one respect, convincing and important. In the first evening, the southern horn, as two observers agreed, changed its form very quickly, that is *in 15 minutes*, so much that the difference between it and the northern was not nearly so striking as before. In the second evening, the air being clearer, and the image excellent, this change was still quicker; for *in 11 minutes, during the observation itself, the end passed very* EVIDENTLY *to the form of a separate point of light.* Supposing both changes to be the same, and produced by the rotation, the alteration to a separate point of light must have happened on the first evening, *at most* 11 minutes later than 6ʰ 40′, when I intermitted my observation; that is, about 6ʰ 51′; because on the second evening it took place in 11 minutes. But on the second evening, when I noticed this striking alteration, I no longer knew the time marked the evening before, and I now noted down 6ʰ 11′. *Consequently this change took place the second time* VERY NEARLY *in 24 hours less 40 minutes; and from these two careful observations alone we may conclude, very probably, the rotation to be nearly* 23 *hours* 20 *minutes*; which agrees extremely well with the approximate period of 23 *hours* 21 *minutes*, which I have deduced from observations of two years, in my circumstantial memoir already quoted.

*Feb.* 28, *from* 10$^b$ 50' *to* 11$^b$ 30', *a. m.* With powers 95, 160, and 209 of the 7-feet SCHR. I found no spot, and both horns perfectly alike ; the light decreasing toward the boundary of illumination extremely plain, and the terminating arch of both horns, but *particularly of the southern,* rather unequal and knotty.

At 3$^b$ 10' to 26', with 160, 209, 370, and 632, a fine image ; the decreasing light seemed at the boundary of illumination to mix itself with the colour of the heavens, becoming equally faint. *Both horns alike oval.*

At 4$^b$ 36', the same.

    5$^b$ 4', no difference.

    6', still the same.

    7', the southern horn began to acquire a pointed shape.

    9', it appeared already pointed ; the northern blunt as before.

    11', the southern exhibited the same appearance as both evenings before ; and I likewise perceived something darker making an impression into it.

    17', Venus behind clouds.

    19', through light clouds her southern horn was perceived to be pointed in comparison with the northern.

    37', the same in some clear intervals. The northern horn appeared always blunt.

That the decrease of light toward the boundary of illumi-nation, whereby that part of the disc becomes extremely dim, is no deception, appeared now evidently ; for whilst the planet faintly glimmered through the clouds, I could often see only $\frac{2}{3}$ of her illumined part, reckoning from the outer edge, and sometimes only half.

$5^b$ $55'$. Venus shining out for a short time between the clouds, the same appearance with full certainty ; and I re-marked also again a slight darker indentation at the southern horn ; *but the scene was by no means so striking as both evenings before.*

$5^b$ $59'$, the appearance changed ; and

$6^b$ $7'$, this was found to be confirmed ; but I could not with certainty discover a separate point of light ; sometimes, how-ever, though but seldom, there seemed a glimpse of it at the southern horn. Immediately afterwards Venus was covered with clouds.

$6^h$ $30'$ to $6^h$ $45'$ Venus shining in a clear sky, her southern horn was again, as at $5^h$ $6'$, rounded exactly like the northern ; and with powers 160, 209, and 370, and a distinct image, I found no trace of a separate point of light. Comparing this third observation with the two former ones, it agrees very well to the minute ; for now the southern horn had nearly the same appearance of being like the northern, at $6^h$ $30'$, as it had the preceding evening at $7^h$ $12'$, and therefore 42 minutes earlier ; but in general it was evident that the appearance remained no longer exactly the same as on the two evenings before ; and this difference may be easily explained by the very probable supposition of a libration, and that it is not a single mountain which occasions the appearance, but a considerable ridge, with many high points: moreover the clearing up or thickening of the atmosphere of Venus, which according to my former observations is pretty dense, and the effects of refraction, may have a considerable influence on such phænomena. Whoever has frequently observed in the moon the very striking variety in the projections of the high ranges of mountains at her edge, namely, Leibnitz, Doerfel, or d'Alembert, will more readily comprehend such effects of a libration.

  *The 1st, 2d, and 3d of March,* bad stormy weather.

  *The 4th, 6ʰ to 6ᵇ 30′, p. m.* with 160 of the 7-feet SCHR. the image being extremely fine, I found *both horns equally rounded, without any difference.*

At 7ᵇ, the same. But at this time there appeared, in the enlightened part, a slight nebulous shade, which, as is expressed in fig. 6, extended to the boundary of light. At 6ʰ, in the bright twilight, I had not remarked it; and I suspected it to be a sort of dazzling, though the image appeared uncommonly soft and distinct. The bad weather which came on soon after did not allow me to apply other magnifying powers and telescopes.

*March 5,* at 4ʰ 25′ to 35′ *p. m.* with the same power, I found the northern horn still rounded, and the southern somewhat *pointed, but not strikingly so.*

At 4ʰ 40′, with a power of 200, the same; and moreover a weak shadow was again perceived on the planet. *So likewise with 288 very distinct, and then with 370 extremely certain; but on the whole it was not striking;* for the southern horn also appeared somewhat roundish, and probably another person less accustomed to such observations, would not have remarked it.

At $5^b$ 45', the atmosphere being less clear, it was doubtful; and at

$6^b$ 35', it was quite certain that both horns appeared equally rounded, without any difference. I found neither spot nor glimmering.

From the 6th to the 10th of March, the learned and worthy Dr. CHLADNI, inventor of the euphon, observed with me ; and having ascertained, by careful comparison, the extreme goodness of my reflectors, can bear witness of it.

*March 6th,* cloudy.

*March 7th,* noon and afternoon cloudy.

At $6^b$ *in the evening* I found, with the 7-feet SCHR. and magnifying powers from 160 almost to 400, both horns constantly the same, without any difference. So they appeared to me also with the 13-feet reflector ; and with both instruments to Dr. CHLADNI.

The 8th March, at noon, the image of Venus but seldom appeared fully distinct. In the intervening moments of greater distinctness, Dr. CHLADNI remarked, *that though* both horns were roundish, yet the northern *was rather more pointed than the southern.* Afterwards I found the same thing. In the afternoon cloudy.

*From* $6^b$ *to* $7^b$ *in the evening,* with 95 to 288 magnifying power, I found *both horns equally round,* and no spot or any thing remarkable, though Venus did not appear perfectly distinct.

*March 9th,* $6^b$ 15', *p. m.* Venus being near her greatest eastern elongation, both horns appeared pretty pointed, with a power of 250, and a fine soft image ; they were also *both alike,* but with the slight difference, that close to the southern horn *a very minute particle projected, which seemed to be rather sepa-rated from the rest of the enlightened part.*

At 8ᵇ 2′, *the air being clear, a projecting inequality shewed it-self with certainty at the southern horn, as is represented in fig.* 7, (Tab. XIII.) *at b.* It was found the same with 288 of the 13-feet.

As our own atmosphere was then very clear, that of Venus also seemed to be purer than usual ; for with both reflectors, and particularly with the 13-feet, Dr. Chladni, *as well as my-self,* enjoyed a magnificent view of the arch of illumination, which seldom presents itself so well to the eye, *the image being uncommonly clear and distinct. To both of us the boundary of il-lumination, toward which the light became very dim, appeared* (be it ever so much contradicted) *not only nebulous, and not sharply terminated, though sensibly sharper than usual, but also very evidently unequal and rugged, with faint shades between,* as I have often seen it, but never so plainly. In truth, the ap-pearance, as each declared, was *very like* the image of the moon at the time of her quadratures, only that the boundary of light was sensibly less sharp, and the faint shadows between were not almost black, but in some measure like the dark spots of the moon's surface, grey, yet darker than the other parts. This instructive observation remains still before my eyes. So delicate a picture of nature cannot well be drawn, however we both made cursory delineations of it, from which fig. 7. is co-pied : but at the boundary of light, soft grey shadows must be imagined, traced into the interstices at *a, b, c, d, e, f, g.*

*March* 11th, *from* 6ᵇ 10′ *to* 45′, *p. m.* the weather having cleared up after snow, I found *no striking difference of the horns,* with powers of 209, 288, and 370, and a distinct image ; how-ever, the southern appeared rather less pointed, which was occasioned *by a* VERY FINE *glimmering pointed line of light, that ran on from the horn not far into the dark side, as at a, fig. 8. and was visible with all magnifying powers. I saw this line of light equally, whether I observed with the whole aperture, or covered a considerable part of it.*

It would be singular indeed, and most discouraging for all such observations, if so many appearances, agreeing together, and viewed with every precaution, should be merely deception, particularly as they usually and principally occurred only at the southern horn, without any reason that could be assigned if it be thought a fallacy. But if there be no deception, it follows incontrovertibly, that the surface of the southern hemisphere of Venus, like that of the moon, has the most and greatest inequalities.

March 12th, $6^h$ 15′ to 30′ *p. m.* no kind of difference in the horns, no spot, or any other unusual appearance, could be seen with a power of 209.

At $8^h$, the same.

But on the *13th of March, from $11^b$ to $11^b$ 20′ a. m.* I perceived, with the *same* magnifying power, *a very evident and remarkable difference. The northern horn appeared pointed, but the southern was rounded, with a very small knot close upon it to the south,* as at *a,* fig. 9. Thus I saw it with 160 and 288 magnifying powers ; and I even distinguished it with 95, though this was too small a power for so minute an object. On the northern horn I found nothing similar, notwithstanding I compared them repeatedly. Business called me away ; and the atmosphere soon afterwards became cloudy, and continued so all day.

This very remarkable observation is indeed not precisely the

same as those of the 26th and 27th of February : yet the
appearance is very little different from that of the abovementioned days, when the shadow, fig. 4. at length penetrated
quite through, and the separated part was perceived as an in-
sulated bright point. Now if it be considered, that on the 28th
of February, only 24 hours later, this appearance recurred, but
was not exactly the same ; and that when a very extensive
mountainous southern region forms the edge of the planet in
various degrees of obliquity, according to the respective situations of Venus and the earth, the phænomena must naturally
be so diversified ; there cannot be the least doubt, but that the
same southern range of mountains, which occasioned the similar appearances of the 26th, 27th, and 28th of February in the
evening, also produced this of the forenoon about 11 o'clock,
according to the rotation ; especially as no intervening observation contradicts this conclusion. The effect of small differences in the position of planets, may be exemplified from
the late eclipse of the sun on the 5th Sept. 1793, when the
projections of the mountains Leibnitz and Doerfel, bounding
the southern edge, were so different from those of the older
observations, under a similar variety of circumstances. The
abovementioned conclusion with respect to Venus becomes still
more evident and remarkable, from its *agreeing more exactly
than could be expected, according to the circumstances, with the
period of 23 hours 21 minutes, which, in my memoir on the rotation of Venus, I had determined as near the truth :* for on the 27th
of February that appearance took place about 40 minutes earlier
than the evening before ; and the middle of the time when the
southernmost part of the southern horn appeared as a separated point of light (a phænomenon similar to the present),

was by that observation at 6ʰ 29'. *From the 27th February,
1793, 6ᵇ 29' p.m. to the 13th March 11ᵇ a.m. there are 13 days
16 hours 31 minutes, which, with the period of 23 hours and 21
minutes, are resolved into* 14,04 *revolutions, exact to the very inconsiderable fraction of* ₁₀₀⁴; which is so much the more sur-
prising, as no attention could be paid to the inequalities.

*The same day at 6 p.m.* I saw Venus with a power of 160,
very sharp and distinct through thin clouds; and found both
horns again equally pointed, and the much fainter light at the
boundary of illumination very evident. And the weather *on
the 14th of March,* having been bad all day, I saw, together
with my attendant, the same thing on the

*15th of March, at 6ᵇ 30'.* Both horns were then alike, and
there was no spot.

*March 16th, 2ᵇ 15' to 45', both horns equally pointed*; no spot.
To search with the greater certainty whether I could not discover some inequality, I took the 13-feet reflector, and still
found it as before, the image being uncommonly sharp. Thus
one observation gives weight to the other against fallacy.

From the 17th to the 21st of March, variable and cloudy
weather.

*March 21, at 7 in the evening,* with powers 160, 288, and
even 95, of the 7-feet, both horns were pointed, without any
perceptible difference: no spot.

*March 22d, 2ᵇ 35' p.m.* the same.

At 7ʰ in the evening, however, I found *a sensible alteration,*
with 160, 209, 288, and 370 magnifying powers. *The northern
horn constantly appeared, according to fig.* 10, *not pointed as before, but somewhat less obtusely rounded,* whilst *the southern was*
pointed and *projecting a little* beyond the line *of the cusps.*

Between the projecting point and the enlightened side, there was often to be perceived, and equally with all magnifying powers, a light greyish shade, which seemed to divide the point.   Soon after the weather became cloudy.

*March* 23*d*, 6$^b$ 37', the atmosphere having cleared up much, but the air being still not very favourable, I found, with the same magnifying powers, *an exactly similar appearance* ; *but an hour afterwards, the northern horn ran out in the same manner into a point, and projected as far as the southern, so that the phæ- nomena were no longer the same.*   Soon afterwards it became cloudy.

*March* 26*th*, 6$^b$ 10' *p. m.* the weather having cleared up again, I saw *both horns equally pointed*, with the same magni- fying powers.

7$^b$ 30', *the same.*

8$^b$ 15', *also the same.*  This too agrees with the period of ro- tation, according to which the phænomena, observed on the 22d and 23d at 7$^h$ and 6$^h$ 37', could not be visible again at the times here noted down.

*March* 27*th*, 11$^b$ to 11$^b$ 40' *a. m.* *both horns equally pointed,* and, as usual, no spots.  With a reference to my former re- marks, I had proposed to observe Venus every hour through- out the day ;  but it grew cloudy.

*At* 6$^b$ 30' *p. m.* the sky having cleared in the part where Venus was, I found in like manner *both horns* equally pointed.

At 7$^h$ 30' *p. m.* the same.

*March* 28*th*, 10$^b$ *forenoon*, with 160, 209, and 288, *both horns were pointed*, without any striking difference.

11$^b$ 15', with the same, both horns *equally pointed.*

5$^b$ 30', *the same* ; even with a magnifying power of 370 times.

$6^b$ 10′, *just the same.*

*March* 30th, $6^b$ 45′ *p. m.* with 160, both horns uncommonly sharp, and *equally pointed.*

$7^b$ 30′, the same. No spot. Then followed rain and cloudy weather.

*April* 2d, $6^b$ 50′ *p. m.* with power 160 of the 7-feet Schr. it struck me *with uncommon certainty and precision, after so many similar appearances of both horns, that the southern horn b, fig.* 11, *was remarkably slenderer in comparison with the northern, a*; *and that in general the whole southern illuminated part, c, b, d, appeared considerably smaller than the northern, c, a, d.* I tried this phase with 288 and 370, and found it to be *assuredly so*; *and with the same certainty I observed it also repeatedly confirmed with the noble* 13-*feet reflector, till* 8 o'clock. *My attendant, who knew nothing of it, made the same remark, and particularly noticed the irregular form of the arch bounding the illumination,* which, by entering in further from *d* to *e*, than from *d* to *f*, formed a slenderer horn, as often happens with the moon; and also in the same manner in its single parts, the crescent of Venus *appeared uneven, like that of the moon,* although not sharply so, but faintly and undefined. I did not now see the mountains of Venus, by their projection and shadow, as in the moon; but the appearances above described must indisputably have been occasioned by mountainous inequalities. Very often have I perceived similar phases on the moon with my naked eye.

It would be inexplicable, if different eyes, with different excellent telescopes, and various magnifying powers, should have seen for an hour together such an appearance, with equal confidence, and yet the whole be nothing but a fallacy, misleading a careless observer. Did not CASSINI, BIANCHINI, and other observers, surely not deficient in caution, perceive similar phæ-nomena, and draw the same conclusion?

At 8$^h$ 35′, Venus presented not a clear image. She had already passed the pleiades about half a degree, and my hope of seeing perhaps an occultation was frustrated.

10$^h$ 15′. A very instructive observation, by comparison with the preceding. Notwithstanding Venus was got near the horizon, and had some tremulous motion from the fine vapours, the sky being otherwise clear, yet her image was free from false light, and sufficiently distinct, with power 160 of the 7-feet SCHR. a reflector which almost never fails me. *I was quite surprised to perceive most evidently, at the first sight, that the abovementioned remarkable phase had changed as remarkably within 2 hours 15 minutes; and that, even whilst the instrument was screwing to its focus, in all parts of the field, the northern horn a, fig.* 12, *constantly appeared pointed; whereas the more slender point of the southern horn, b, had vanished, and this horn had become rounded, as it was on the 26th, 27th, and 28th of February, and the 13th of March.*

Comparing this observation with those I have here named, it becomes very remarkable and decisive, by confirming my former approximated estimate of the period of rotation. On the days just mentioned I had, at the hours noted down, observed a somewhat similar change in the southern horn, *conformably to such a period of rotation;* but had never seen it again in all the numerous observations I made since the 13th of March, at hours when, according to the rotation, it should not appear. But now it was seen again at 10$^h$ 15′ in the evening. *From* 11$^b$ *in the forenoon of the 13th of March, to the*

*2d of April at* 10$^b$ 15′ *in the evening, there are* 20 *days* 11 *hours and* 15 *minutes, which, with a period of rotation of* 23$^b$ 21′, *divide into* 21,005 *revolutions, exact to the inconsiderable fraction of* $\frac{5}{1000}$.

*April* 3d, 5$^b$ 40′ *p. m.* with 160 and 370 magnifying powers, I found Venus again *irregular* in single parts of the *arch terminating the illumination.* That is, according to fig. 13, (Tab. XIV.) it sunk in somewhat, but very little, at *a,* and between *a* and *b* it protruded out a very little. Both horns, however, were pointed, and no spot could be seen.

*At* 6$^b$ 48′, the boundary of light went in a little at *d* also, according to fig. 14.

*At* 7$^b$ 25′, I found both horns alike pointed, and no striking difference whatever, as the evening before. No spot.

*At* 8$^b$ 10′, the same. No perceptible difference in the horns.

*At* 9$^b$ 50′, I found *the southern horn visibly, though not much, rounded as yesterday. Mr.* TISCHBEIN *saw it so likewise :* but Venus was already too low, and undulated in the vapours, so that we could not reckon on this observation with confidence; yet it agreed with the former.

*April* 4, at 5$^h$ 50′ *p. m.* with a magnifying power of 160 Venus appeared extraordinarily plain and fine, but without spots. The light lost itself in a dim grey at the boundary of illumination, which appeared somewhat uneven, as it did yesterday about the same time, but both horns looked equally sharp.

Without thinking of it in the least, I saw, with a power of 288, *that the southern horn was somewhat slenderer than the day before yesterday ;* and this was confirmed with a power of 370, which shewed me clearly that the smaller form of the right side, according to fig. 15, was occasioned by the boundary of light running in a little more at the right horn.

$6^b$ 25′, I found this repeatedly confirmed with 370.

*At* $7^h$ 5′ *to* 10′, this difference no longer struck my eye; both horns appeared equally pointed.

$8^b$ 24′, the same with 160. Venus was no longer distinct.

*April 5th,* $5^b$ 15′ *p. m.* Both horns indeed sharp, but *all* as it was the evening before, and nothing striking, with 160.

$5^b$ 25′, the same with 288.

$6^b$ 38′, still the same.

$7^b$ 38′ *to* $8^b$ 10′, with both magnifying powers, and afterwards with 136 of the 13-feet, no manner of inequality in the horns. With the greater telescope the decrease of the light to dimness, and the *dim unevenness of the boundary of light*, appeared extraordinarily fine.

$8^b$ 42′. Both horns still equally pointed, with power 160 of the 7-feet. No spot.

$9^b$ 55′, still the same. The planet being now as low as on the 2d and 3d of April about the same time, I tried, by screwing in various ways, whether I could get the southern horn to look somewhat rounded, as it did then, but in vain : both horns were equally pointed.

April 6th, $6^h$ 45′ *p. m.* with 160 of the 7-feet, I found no striking difference, *both horns being equally pointed.* No spot.

$7^b$ 29′, likewise so.

$8^b$ 10′, the same with power 288, and an extremely sharp image.

$8^b$ 45′, the same.

$10^b$ $5'$. In this situation of Venus near the horizon, I tried again, by screwing the small speculum, and moving the image in the field, whether I could give her a false form, similar to that of the roundness of the southern horn on the 2d and 3d of April ; but both horns were, and remained, pointed. Consequently the observations of the 2d and 3d April were no deception, and they agree extremely well with the period of rotation, being the 5th and 6th new repeated proofs of it.

*April 7th, $6^b$ 30'*. With power 160, both horns were equally pointed : no spot, nor any sensible inequality, except the dim faintness of the boundary of light.

$6^b$ 55', with 288, the same.

$7^b$ 15', the same.

$7^b$ 55', still the same.

From the 7th to 12th of April cloudy weather.

*April 12th, $6^b$ 30' p. m.* With the same magnifying power *both horns equally pointed. However, Venus was now become too narrow a crescent for a rounded shape of either horn to be expected.*

$8^b$ 20', the same, without perceptible inequality.

A series of changeable bad weather.

But on *the 22d of April in the evening,* the hour not being marked in the journal, *the southern horn* appeared to be illuminated *only half as broad as the northern.*

*April 23d, $5^b$ 45', till after $6^b$ p. m.* With 160 and 288, Venus was *distinct, and her southern horn again much smaller than the northern, according to fig. 16.*

*But at $10^b$ there appeared no longer any striking difference.* However Venus was already got too low, and I would never advise a careful observer to choose such a time for investigations of this kind.

*April* 30*th,* 7*ᵇ*. *Judging from the outer circle, I found the northern born running out much longer than the southern.* At the same time the southern appeared sensibly smaller: see fig. 17. I leave these remarkable phases to the judgment of the skilful, but to me they seem inexplicable, except from real shadows of an uneven mountainous surface.

*May* 3*d,* 7*ᵇ p. m.* After much rainy weather I saw a similar phase; for though I found both horns, at 7*ᵇ* 30′, without any sensible difference in their length, yet the northern was evidently broader than the southern.

7*ᵇ* 45′, still the same.

8*ᵇ* 25, the southern horn was still somewhat smaller, but only a little.

9*ᵇ* 45′. Venus being now near the horizon, and undulating in the vapours, I could perceive *no difference in the breadth of the horns.*

*May* 6*th,* 5*ᵇ* 50′ *p. m.* with a very distinct image I found both horns perfectly alike.

*May* 8*th,* 8*ᵇ* 15′ *p. m.* the same, but the image indistinct after storms. No spots; but they are not to be expected in these small phases.

I now longed for fair weather, that I might carefully attend to the twilight from the atmosphere of Venus, which I discovered in 1790, as far as should be practicable in the present less favourable circumstances.

*May* 9*th,* 6*ᵇ* 25′ *p. m.* I found, with full certainty, that though both horns were equally long, the southern at *a,* fig. 18, was *scarcely half so broad as the northern* at *b;* and this was confirmed by continued attention to the object.

7*ᵇ* 50′, still nearly the same.

At 8ʰ 20′, on the contrary, the difference was no longer by far so perceptible.*

This day the first traces of Venus's twilight shewed themselves; for the points of the horns appeared to terminate beyond the illuminated hemisphere, in an extremely faint bluish-grey light.

*May* 10th, 6ʰ 40′. A perfectly similar phase. I found, so as to be quite certain of it, the southern horn only half as broad as the northern; but both horns were equally long.

7ʰ 30′, still the same.

8ʰ 15′. With 180, 400, and 560 magnifying powers of the 13-feet reflector, and a distinct image, I found traces of the twilight which could not be mistaken. The light grew dimmer and dimmer to the point of both horns, and at the points was so dim, that it seemed to lose itself in the faint light of the sky. A still finer dimmer trace of light shewed itself twinkling at both sides, on the edge of the dark hemisphere, and including this the two horns comprehended sensibly more than a semicircle; but it was too fine and dim for me to measure its extension.

Even if I had not seen this, I should repeatedly have obtained conviction of the particular density of Venus's atmosphere, by the faint colour of the points of the horns, and of the boundary of illumination.

* It is scarcely necessary to put the reader in mind, that small, undulating, knotty inequalities of the boundary of light, in such observations, must not be taken for true inequalities, or mountains of Venus. In general, these small crescents, as the enlightened part lies obliquely to the eye, are not well suited for observing the true inequalities of the boundary line, or any spots there may happen to be. For such observations, we should be assiduous in attending to the planet, about the time of its greatest distance from the sun.

U

*In this reflector likewise, as well as in that of 7-feet, the south-ern horn appeared sensibly smaller than the northern.*

*May 12th and 13th,* I perceived again traces of the twi-light of Venus; but the stormy state of the air rendered it too bad for such nice observations.

*May 16th, after sunset to* 8ʰ 40′, I had, for the third time, the pleasure of observing this crepuscular light of Venus's at-mosphere, with the 13-feet reflector. Although the circum-stances were not by far so favourable for such observations as when I discovered it in the year 1790, and the luminous ap-pearance therefore came to the eye sensibly weaker and more indistinct than at that time, yet all was confirmed; and in this observation I thought it worth remarking, that the dim cre-puscular light seemed to extend sensibly further on the south-ern than on the northern horn, though this might easily be a deception.

*May 19th,* after sunset, the light now coming to the eye sensibly clearer, I found the circumstance just noticed to be again the same, with 97 of the 7-feet HERSCH. and 136 of the 13-feet.

Hitherto the circumstances had not been favourable enough for a repetition of the measurement, and therefore I was eager for a better observation.

But *May 20th,* Venus was covered with clouds. However, at length I succeeded in a measurement,

*May 21st, at* 8ʰ 30′, *p. m.* six days before the inferior con-junction, and consequently just the same time as in the year 1790. Venus being rather too low for the 13-feet, and for the 7-feet HERSCH. I employed the 7-feet SCHR.; and found the crepuscular light beautiful, and sufficiently distinct. *It extended,* according to fig. 19, (Tab. XV.) *from the proper points of the horns a, b, a considerable way, on the edge of the dark hemisphere, to d, e; and equally far on both sides, having the appearance of a very dim, constantly decreasing light.* But I must remark, that in the present more unfavourable situation of Venus, it did not affect the eye as a bluish-grey light, which was its appearance March 12, 1790, but only as a dim grey light.

According to my usual projection-measure,* in which each decimal line of the projection table is equal to 4″ of space, I found the apparent diameter of the planet *a c b,* after repeated trials, = 15 lines = 60″; the projection of the crepuscular light running into the dark hemisphere *a d, b e* = 25 lines = 10″, *and fully so, being rather more than less.*

As the crepuscular light could be distinguished from that of the points of the horns, by its sensibly fainter colour, I was able to measure it from the points. But in order to know with certainty whether I had taken the true termination of the

---

* In the year 1790, as well as in 1793, I measured this crepuscular light with a projection-machine, which is nothing more than a very simple projection-micrometer, useful in many cases, both by day and night: it gives, for all magnifying powers, the measure of the projected object immediately in minutes and seconds of space, without the necessity of first measuring a fundamental line. I contrived it for my purpose of a selenotopography, and constructed it myself. After an experience of many years, I certainly would not lay it aside, in most cases, it being so quick in the use. I have described it, in all its simplicity, in my *"Beyträge zu den neuesten astronomischen Entdeckungen,"* p. 210, where the *older* lamp-micrometer of the worthy Dr. HERSCHEL is *also described* BEFORE, *p.* 138, with which this machine may be compared. It has never made pretensions to be a new invention, because projection-micrometers of many kinds, for example, accompanying microscopes, have long been known. I remember with pleasure that, even in the year 1778, the window frames were my projection-micrometer, on which I determined the proportion of magnifying powers to one another.

horns a, b, for the foundation of my measurement, I measured likewise the two lines a f, b g, perpendicular on the line of the cusps. I found the northern side $f a = 8$ lines $= 32''$, the southern only FULLY 7 to $7\frac{1}{2}$ lines; mean, $7,25 = 29''$; consequently both sides together $f a + b g = 61''$, therefore the mean of each side $= 30'',5$; but if the southern side b g be put $= 7,5$ lines, the mean will be fully $= 31''$; so that, as the semidiameter, according to the first measurement, could amount only to $30''$, I probably observed the cusps as projecting $0'',5$, and perhaps something more, beyond their proper line;* and consequently *the projection of the crepuscular light, which extended into the dark hemisphere, was certainly and at least as* 1 : 6 *in proportion to the apparent diameter.*

My success in this measurement was the more lucky, as on the 22d *of May* Venus could no longer be discerned, though the air was clear.

These are my late observations, made about the time of the greatest eastern elongation, in the year 1793; and continued three months to the inferior conjunction. Under my present circumstances, I hope to be excused for giving them with such prolixity; but I should quite weary the reader, were I now to lay before him likewise my further observations, continued to the last western elongation; which, therefore, I shall rather reserve to another occasion, especially as they contain little that is interesting.

However, I must not leave unnoticed some conclusions, remarks, and explanations, which are deducible from these observations; and which have for their object, partly the moun-

---

* The remarks and computations that follow hereafter, will shew that the penumbra was probably included in the measurement.

tainous inequalities and period of rotation which I formerly discovered, and partly the atmosphere and crepuscule of Venus.

## I. *Remarks on the Mountains and Rotation of Venus.*

1. As I have already said, I gave, in the memoir on this subject published last summer, only those observations which particularly belonged to the object, out of a *very great number* that I had made during 13 years; and I omitted the rest, because otherwise they would have amounted to a volume alone. Now with regard to those I have communicated, and which shew the real existence of considerable mountains, as well as an approximate determination of the rotation, the respectable author of the paper against me has not observed the planet Venus *once at the same time,* which might easily be the case in only 38 observations, that are adduced from a period of 15 years. But in the numerous remaining observations, I saw neither mountains, inequalities, nor spots, any more than the author; and I doubt not, that among these observations I should find many which were made at the same times as when he observed. The same holds good

2. With respect to the new observations for three months, here communicated, which amount to more than 100, and were made at various hours, and on different days. Of the 25 adduced by my opponent, there are *only* 4 made nearly at the same hour, which is the chief circumstance ; and not only *in all these, but likewise in very many other observations, I saw, exactly as he did, no spot, and both horns like each other* : so that of all his observations, not one contradicts mine. And yet it would

not be a decisive contradiction, if some observations made at
the same time by another person, were in opposition (though
that is not the case) *to so many of mine, made in various ways, yet
agreeing together;* because, when fallacy of vision is in question, it may always be doubted which of the two observers is
deceived; since this depends on the goodness of sight and of
instruments, but much more on care and caution.*

I confess impartially, that, before reading the observations
contained in my two memoirs, I should have formed the same
judgment from those of the abovementioned author that he
has done; and on that account his paper is highly valuable to
me, as leading to a more scrupulous examination of new truths.

3. However, that which these new observations, here communicated, clear up and confirm, in correspondence with my
older ones, on the mountains and rotation, is, that the planet
Venus has very considerable mountains and elevated ridges;
and indeed the most and the highest in her southern hemisphere. This appears

(*a*) From the observations of the boundary of illumination,
which is not sharply terminated, and seems formed of light
and greyish shadow indistinctly intermingled. This is chiefly
to be perceived only about the time of the greatest elongation,

---

* By comparing the respective times of the two observers, it appears that both of
them viewed the planet on the 4th of April at $7^h$ $30'$, *p. m.*; the 5th April at $6^h$ $38'$,
the 6th April at $6^h$ $38'$, and the 7th at $7^h$ $15'$, *exactly at the same time, and saw exactly the same appearance.* The comparison of these observations is the more instructive, because I did not, like my opponent, observe Venus only once, but as often
as was possible each day, and *at other times, on the same days, found evident changes;*
for this shews plainly enough, that whoever wishes to see the same, and as much, in
Venus, must observe with equal industry, and on each day as many hours as possible,
with the same care.

when the eye looks perpendicularly through the dense atmosphere of Venus, and by no means in the small crescent form of light, when the lines of vision are much longer and more oblique through that atmosphere: it is in the former position of the planet alone that it can be seen distinctly, but even then not always equally so. One of the finest scenes of this kind was afforded (for example) by the observation I have adduced of the 9th, when Dr. CHLADNI viewed the planet with me. A less striking inequality, though perfectly certain, was discovered by my learned friend Dr. OLBERS, July 31, 1793, at $11^h$ $5'$ in the forenoon, which we both observed and delineated in the same place, and exactly similar, after we had been observing since $3^h$ $15'$ in the morning, but till that time saw no inequality. Were these small indentations or darker places merely atmospherical, no reason can be perceived why they should shew themselves only in the boundary of illumination, and not in the other enlightened parts also.

(b) The same thing appears, moreover, from the irregular form which the arch bounding the illumination sometimes assumes, and from the phænomenon thence arising of the much smaller size of one horn, and particularly the southern, in the crescent-shaped phases of the planet; as is shewn, on the same grounds, by the observations contained in my former memoir on the rotation.

Were these observations, as is alleged of the rest, nothing but fallacy, I should wish to know the reason, why that deception happens only sometimes, continues only some hours, and almost always takes place on the southern horn only, very seldom on the northern. Whoever compares together the observations of this kind contained in my memoir on the rotation, to which I have referred, § 12 to 23, will find 14 in which the southern horn appeared much smaller than the northern, but only one or two instances of the opposite phænomenon. And, if it were merely deception, why does the smaller horn, when the planet is seen through light clouds, always disappear sooner than the broader one, and become visible again later? (See § 12. No. 4. of the memoir.)

It further appears likewise,

(c) From the observation, that *sometimes, though much seldomer*, one horn, and particularly the southern, is seen rounded *about the time of the elongations*, but the other pointed. And by this very circumstance chiefly is

4. The period of rotation, which I had concluded to be nearly 23$^h$ 21', confirmed and rendered evident by the new observations given above.

Having already explained this curious circumstance when the observations themselves were stated, I will here only make the following remarks.

(a) If the very remarkable observations of the 26th, 27th, 28th February, 13th March, 2d and 3d April, when the southern horn appeared rounded, but the northern pointed, are compared together, the abovementioned period will be found to suit them all, during an interval of 37 days, as exactly as can possibly be expected, and indeed to very inconsiderable fractions. If, on the other hand, they are compared with the older observations of this phænomenon, namely, those of 28th December, 1789, 31st January, 1790,* the 25th, 27th, and 30th

* See *Selen. Fragm.* § 522.

Dec. 1791, and the 11th Jan. 1792,* the differences are more considerable. Thus, for example, from the most distant observation, on the 28th Dec. 1789, at $5^h$ $p. m.$ to the 27th Feb. 1793, at $6^h$ 41′ $p. m.$ are 1157 days 1 hour and 41 minutes, which dividing into 1189,28 revolutions, might occasion some doubt. But,

($\alpha$) In each separate period, several observations correspond as well as can be desired:

($\beta$) The period is only assigned nearly, but the interval of more than four years is very long, so that an error of seconds may occasion such an excess; and accordingly the abovementioned time would divide even with a period of $23^h$ 21′ 19″: and

($\gamma$) In such computations, no regard is paid to the inequalities of the planet, nor to the middle of the duration of the phænomenon : wherefore so considerable a length of time can never be divided exactly by the period ; as my observations of the rotation of Jupiter likewise could not, under similar circumstances, though the period of that rotation is sufficiently well known.†

($\delta$) A like doubt might arise from the phænomenon being sometimes not at all or very doubtfully perceived, about the times of the greatest elongations, even at the hours when it was to be expected, according to the period. Hitherto, however, during more than four years, only three instances of this have occurred to me; which were in the years 1790 and 1791, and about the time of the late western elongation, in August, 1793; in which last I only *twice* perceived *barely a trace* of a somewhat rounded form on the southern horn. Moreover, as often

---

* See *Beob. über die sehr beträchtlichen Gebirge und Rotation der Venus*, § 26 to 30.

† *Beyträge zu den neuesten Astronom. Entd.* p. 1 to 138.

happens, the weather was not always favourable; and besides, the observations already communicated contain sufficiently evident marks of a libration, whence such cases may be easily explained. So, for example, the mountainous ridges of the moon's southern edge, *Leibnitz* and *Doerfel*, do not shew themselves quite clearly at each rotation, but only sometimes arrive at their full projection.

(*c*) But the very circumstance, that during more than four years, in so great a number of observations, I have perceived this phænomenon only ELEVEN TIMES with perfect certainty, and only a few other times uncertainly, and that in all the intervals I have expected it in vain, notwithstanding my frequent wishes, seems alone to shew, evidently enough, that I cannot have been deceived; especially as those appearances have been seen, with various magnifying powers of different telescopes, and in several instances with different eyes, perfectly alike, and with full certainty; and it is not reconcileable to our understanding, how such a fallacy should, at different times, always preserve one and the same period.

The following example, which I here take an opportunity of adducing as remarkable, may shew how cautious we ought to be, in drawing conclusions from our own observations, against the truth of those made by others. Jan. 5, I reviewed with the 13 and 25-feet reflectors the Mare Crisium (HEVEL. Palus Mæotis) in the moon, and made some observations. The following day, Dr. OLBERS of Bremen, who now pursues his observations with an extremely good 5-feet DOLLOND of $3\frac{3}{4}$ inches aperture, mentioned to me, that he had discovered *the preceding evening*, in the *Mare Crisium*, between Picard and Auzout, two small craters in the grey plain, which were both wanting in my topographical charts; and about which, therefore, the question might arise, whether they were not newly produced?—I had seen nothing of them with my more powerful instruments. Again, on the 6th, I examined the part of the surface which he had exactly pointed out, with powers 186 and 300 of the 18-feet reflector, and found nothing. The 17th, I looked for them with the 7-feet, in vain. I did the same on the 3d February, with 179 of the 25-feet, and likewise on the 6th, but found not these craters. Hence I might have concluded, with probability, that the learned observer had been exposed to some deception; and perhaps I should have been believed. *And yet Dr.* OLBERS *was perfectly in the right.* On the 6th of March, I readily found the largest of these two craters, without seeking for it long, and saw it *uncommonly sharp and clear*, with 160 and 280 of the 7-feet SCHR. It is very nearly as big as a crater which I discovered last year, lying also in the plain, between the eastern bounding mountains, where they break down; it is surrounded with a broad, and proportionably flatter, annular elevation, of little brightness; it appears to be uncommonly deep, in proportion to its breadth; and if a straight line be conceived, running from Picard* towards the middle of the southern boundary mountain, which projects inward in the shape of a wedge, it lies on this line about $\frac{2}{3}$ distant from Picard. As I have examined this tract of the Mare Crisium very often, and under the most favourable angles of illumination,† in searching for the veins of mountains, or the flat mountainous layers to be found there, but *never perceived the slightest trace* of these craters, the

* See Tab. VI. of the *Selenotop. Fragmente.*
† See Tab. XXXIII. XXXIV. and XXXV. of the same work; and § 355 to 397.

observation of Dr. Olbers is certainly not unimportant, and it will on occasion be further explained.

If any astronomer shall think it worth the trouble to observe *Venus, not barely now and then, at whatever time of the day it may be, but continually, with the same persevering zeal, and when the weather is favourable almost hourly, about the time of her greatest distance from the sun,* I am convinced that he will certainly perceive the rare phænomenon in question, just as well as I have done. If, contrary to all reasons which hitherto appear, I should hereafter be convinced that I was deceived, I would myself, willingly and impartially, bring the offering to truth; and so much the more readily, as no indirect views have ever led me on, but I have been actuated solely by an irresistible impulse to observe ; and because I certainly shall never have reason to be ashamed of the observations I have laid before the world, which have always conducted me to new truths.

II. *Further Explanation and Correspondence of Computations of the Twilight, together with Remarks on the other Properties of the Atmosphere of Venus.* *

As the celebrated author of the paper so often mentioned, " on the planet Venus," though he confirmed my discovery of the twilight of Venus's atmosphere, *yet represents the computation of it, p. 16 and 17,† as not demonstrated, and positively as very inaccurate,* which may, without any foundation, be injurious to the truth, it becomes my duty to give some explana-

---

* Many of the explanations and remarks in this section come from Dr. Olbers of Bremen, who, at my request, kindly undertook not only to examine the old computation, but also to compare the calculations deducible from the new observations.

† P. 214 and 215, Phil. Trans. for 1793.

tions and remarks, that persons skilled in those matters may be better able to form a right judgment of my new computation, which agrees excellently with the old one; and at the same time may determine, whether there be inaccuracy and error, and on whose side it lies.

1. The first objection is concerning the apparent diameter of the sun, as seen from Venus, which I have assumed at 44', in the computation of the penumbra, smaller, it is alleged, than I ought to have taken it.

M. DE LA LANDE puts the diameter of the sun in the apogee $= 31' \, 31'' \doteq 1891''$. Now the apparent diameter seen from Venus

$$= \frac{1891'' \times dist. \; \odot lis \; in \; apog.}{dist. \; Ven. \; a \; sole};$$

consequently,

$$\log. \; 1891 \;\; = 3{,}276692$$
$$\log. \; dist. \; \odot \; = 0{,}007231$$

$$\overline{\phantom{xxxx}3{,}283923}$$

| | | |
|---|---|---|
| log. of the distance of Venus *in aphel.* | - | 9,862318 |
| log. of the distance of Venus *in peribel.* | - | 9,856337 |

| | | |
|---|---|---|
| log. of the diameter of the sun *in aphel.* | | 3,421605 |
| log. of the diameter of the sun *in peribel.* | - | 3,427586 |
| Diameter *in aphel.* | $= 2640'',0 = 44',0$ | |
| Diameter *in peribel.* | $= 2676'',6 = 44' \; 36'',6$ | |

But if the assumed diameter of the sun in the apogee 1891'' be corrected for the irradiation, which may be put $= 6''$ (DE LA LANDE *Astron.* § 1388), we have the diameter of the sun seen from Venus

*in aphel.* = 43′ 51″,6

*in peribel.* = 44′ 28″,1

I really do not see, therefore, how the diameter of the sun seen from Venus could be expressed generally, and with respect to every part of her orbit, more accurately, than as 44′, the quantity taken for the calculation. And indeed equally unimportant, must be considered

2. The remark on my computation of the penumbra. The sense of the note on that subject, which I have given, p. 313 (Phil. Trans. for 1792), is plain enough, that, as the sun is seen in Venus under an angle of 44′, the penumbra, assuming the diameter of Venus = 60″, can amount only to 0″,38 in the middle of her disc; but that as Venus, when her diameter is so large, can only appear under the phase of a crescent, the penumbra can scarcely amount to $\frac{1}{10}$ of a second in the perpendicular diameter on the line of the cusps. Instead of 0″,38, or still more accurately 0″,384, by an error of writing or computation 0″36 was set down: but what does this inconsiderable difference, *of* $\frac{1}{50}$ *sec.* impede in the conclusion, *that the penumbra at the boundary of light on the disc, or in the perpendicular direction on the line of the horns, is imperceptible?* and how could so unimportant a matter deserve the least notice?

3. *With respect to the twilight itself of Venus's atmosphere, and the computation of it,* the paper in question contains, p. 16 and 17, three objections: (*a*) that I had *overlooked the penumbra,* which, in the projection I have given of the crepuscule 15° 19′ is said to amount to *more than* 2°⅓, or, as this error of computation was corrected in my copy, to 1° 11′ 47″,6; (*b*) that my 7-feet speculum must be tarnished, *because I have measured the*

*projected extent* TOO SMALL; and (*c*) that *my calculations are so full of inaccuracies, that it would be necessary to go over them again, and compare them* EXACTLY WITH THOSE MADE BY MY OPPONENT.

It requires, indeed, little examination to perceive, that all these objections are groundless.

(*a*) That I did pay attention to the penumbra, my paper " on the atmosphere of Venus" shews plainly enough; and it is readily to be conceived, that the points of the horns, illuminated by refraction and penumbra, must project beyond the enlightened semicircle into the dark side. And it would also be easy to shew, how the points must project more beyond the enlightened semicircle, in proportion as the phase of Venus is that of a sharper crescent; with regard to which, I will hereafter determine, more accurately than my opponent has done, how much the projecting excess of the arch must be. But the author has not considered *that, in my way of making the measurement, it was quite unnecessary to take the penumbra into the computation* ; for I measured the faint light, of a bluish-grey colour, which ran on along the edge of the dark hemisphere, according to fig. 20, (where A D indicates a diameter of Venus, parallel to the line of the horns) not, as he did 3 years after, from A, but only from B (the extreme visible point of the horn, still faintly illuminated by refraction and the diameter of the sun) to C ; *and consequently I had, by the observation itself, already deducted the penumbra.* It is indeed possible, that at B and E, where the penumbra seemed to me to terminate, it yet might not be quite at an end ; but the excess must be indefinitely small, since the *whole* projection of the penumbra, from A to B, and from D to E, could not, by my calculation, amount to more than 0,63 second.

In general, such accuracy of computation avails nothing, *because the observations and measurements of such a very faint and always decreasing light, cannot be so very exact.* This is particularly shewn, and strikingly enough, in the two measurements of 20th May, 1793, given in the paper of my opponent; where the projection of this crepuscular light, taking the apparent diameter of Venus = 60″, was one time 12″,5, and the other time only 7″,7. And so much the more unimportant is it in the result of my calculation, that I assumed the crepuscular light as having been measured from A. But that in my way of measuring, in which the penumbra is abstracted by the observation itself, I have been happier and more accurate, is testified by the computations to be given presently of my two measurements of the years 1790 and 1793, which were made under different circumstances, and yet correspond uncommonly well.

(*b*) The second objection, *that I have measured the projection of the twilight too small,* is equally unfounded ; for

(*a*) The projection found by the author must properly be somewhat larger than mine, because he did not, like me, measure the magnitude B C, but A C, fig. 20; and

(*β*) It will appear from the following computations, that I have found it AT LEAST AS LARGE as he did, without reckoning in the difference from A to B. He did not consider, that three years before I had observed under other circumstances, *which must make the extent of the crepuscule appear less;* and in general I do not perceive how he can form such a judgment from his two measurements, which differ from one another so very much as $\frac{1}{3}$ of the whole magnitude. I can also assure him, that the 7-feet speculum, which I obtained in the year 1786 by his friendly kindness, has continued always so precious to me, that I have kept it in perfectly good condition to the present time.

As to

(*c*) the objection, *that my calculation abounds with inaccuracies,* it is indeed true, that the observation of March 12th, 1790, was not rigorously computed, yet *its exactness was carried much further than is necessary in observations of this kind;* for no one will comprehend the use of a scholastic computation to seconds and decimal parts of seconds, when the observations themselves leave an uncertainty of many minutes. However, to remove all doubt in this respect, and to save the author the trouble of a further careful comparison with his two measurements, I will here not only repeat the calculation in all its rigour, but also add the new one for my second measurement of the 21st May, 1793, and compare both together, as well as with that of my opponent.

(*α*) *Calculation of my observation of the 12th March 1790,* $6^{\mathrm{h}}$ o', *p. m.*

The time of this observation may be taken, without scruple, as $6^{\mathrm{h}}$ o' mean Paris time; for it was made after 6 o'clock at Lilienthal. The equation of time amounts to 10', and the difference of meridians to 26'; therefore, if the observation had been made exactly at six, this would be $5^{\mathrm{h}}$ 44' mean time at Paris.

Now, according to the latest tables by M. DE LA LANDE, we have, for that moment,

heliocentr. long. of Venus        -        $= 5^s\ 18^\circ\ 41'\ 53''$
long. of the earth                 -        $= 5\ 22\ 22\ 45$

difference        -        -        $=\quad 3\ 40\ 52$
heliocentr. latit. of Venus        $=\quad 3\ 23\ 6$

therefore,

log. cos. $3^\circ\ 40'\ 52''$        $-$        9,9991030
log. cos. $3\ 23\ 6$        $-$        9,9992416

9,9983446
angle at the sun        -        $=\quad 4^\circ\ 59'\ 58''$
sum of the other angles        -        $=\ 175\ \ 0\ \ 2$
half sum        -        -        $=\ 87\ 30\ \ 1$
log. of the distance of the earth from the sun $=\ \ 9,997766$
log. of the distance of Venus from the sun $=\ \ 9,857040$

log. tang.        -        10,140726
$= 54^\circ\ 7'\ 28''$
subtract        45   0   0

remain        9   7   28
log. tang. $9^\circ\ 7'\ 28''$        -        -        9,205777
log. tang. $87\ 30\ 1$        -        -        11,359955

log. tang.        10,565732
half difference        -        -        $=\ 74^\circ\ 47'\ 42''$
half sum        -        -        $=\ 87\ 30\ 1$

angle at Venus        -        $=\ 162\ 17\ 43$

angle at the earth               =  12° 42′ 19″
compl. of the angle at Venus =  17  42  17
Now the crepuscular light of Venus, the measure being considered as a chord, extended 15° 19′;, then consequently it is,

log. sin. 15° 19′  0″  -  9,421857
log. sin. 17  42  17   -  9,483033

$$\overline{\qquad\qquad\qquad}$$

log. sin.    8,904890
= 4° 36′ 28″

To so much, therefore, amounts the arch of a great circle,
over which the crepuscule of Venus's atmosphere extends, as
far as it can be distinguished on our earth, under favourable
circumstances. According to my former computation, it came
to 4° 38′ 30″ : *wherefore the whole difference, certainly very inconsiderable to be given as an instance of inaccuracy, amounts* ONLY
TO 2 MINUTES; and it is surely quite superfluous to include
seconds in a calculation, which, from the circumstances of the
observation, can only be depended on to several minutes.

If it be wished to take this opportunity of determining the
arch, how far the points of the horns project on account of the
apparent diameter of the sun seen from Venus, put the semidiameter of the sun seen from the earth at the abovementioned
time, deducting 3″ for irradiation, = 16′ 3″,3 = 963″,3

log. 963,3      -     2,983762
log. dist. ☉      = 9,997766

$$\overline{\qquad\qquad\qquad}$$

2,981528
log. dist. ♀ a ☉      9,857040

$$\overline{\qquad\qquad\qquad}$$

Y 2

l. sin. d. ⊙ *ex* ♀ -    3,124488 = 1332",0 = 22' 12"
l. sin. 17° 42' 17"    9,483033

3,641455 = 4379",8 = 1° 12' 59",8.

This is the quantity which, in the paper of my opponent, was erroneously stated at more than $2°\frac{1}{3}$, because the diameter was taken instead of the semidiameter; it was afterwards corrected to 1° 11' 47",6; but it is highly probable, that the points of the horns project still further, *on account of refraction.* However, as we do not know the quantity of the horizontal refraction on Venus, this cannot be ascertained with any certainty. It is sufficient for me, that I measured the crepuscular arch from the point where the extremity of the horn seemed to end in my instrument, and to my eye.

(β) *Calculation of my late observation of the 21st May, 1793.*

The time falls on 8ʰ 0′, mean Paris time ; and therefore we have,

long. of the earth    = 8ˢ 1° 5' 55"
heliocentr. long. of ♀ = 7 27 13 27

difference    =    3 52 8
lat. of Venus    -    =    1 1 37,6
log. cos. 3° 52' 8"    -    9,9990091
log. cos. 1 1 37    -    9,9999302

log. cos. of the angle    -    9,9989393
angle at the sun    =    4° 0' 10"
sum of the other angles    = 175 59 50
half sum    .=    87 59 55

log. dist. of the earth from ☉ = 0,005628
log. dist. of Venus from ☉ = 9,860276

log. tang. 10,145352
= 54° 24' 50"
subtract 45  0  0

remain  9 24 50

log. tang.  9° 24' 50"  –   9,219579
log. tang. 87 59 55   –   11,456615

log. tang. 10,676194
half difference  =  78° '5 53"
half sum  –  =  87 59 55

angle at Venus  –  = 166  5 48
angle at the earth  =  9 54  2
compl. of the angle at ♀ =  13 54 12

Now, as I have stated above, the projected extent of the twilight measured 10", putting the semidiameter of Venus = 30";
and, as I then measured that extension perpendicularly on the line of the cusps, these 10" may be considered as a sine, and so the arch will amount to 19° 28'. Then

log. sin. 13° 54' 12"   9,380725
log. sin. 19  28  –   9,522781

8,903506

= 4° 35' 34"

To so much, therefore, amounts, according to my *second*

observation, the arch of a great circle, over which the twilight
of Venus's atmosphere extends, as far as we can discern it under favourable circumstances, and which we may put in comparison with our common twilight.

If this result be compared with that of my older observation of the 12th March, 1790, which was 4° 36′ 28″, it will be
seen *that the two agree much more nearly than could have been
expected in such delicate observations, namely, to the very inconsiderable difference of one minute;* and this is the more striking,
as, according to the different situations of Venus, and the modifications of our own atmosphere, this crepuscular light is not
likely to be ever observed, at different times, exactly of the
same extent.

Great, however, as this agreement is, I am far from regard-
ing it as any thing but a lucky accident. Whoever considers
the manner of measuring, and the nature of the observed object, will be easily convinced, that we can never determine
quite exactly the length of the twilight of Venus. The most
accurate measurements of this kind admit errors of $\frac{1}{2}''$ in the
projected extension ; and this $\frac{1}{2}''$ alone would amount nearly
to $\frac{1}{4}°$ in the computed arch of the great circle. Moreover the
crepuscular light gradually decreases, and I only pretend to
shew how far it continued visible, in my observation, with my
eyes and instruments, under the state of the atmospheres of
Venus and the earth at that time : the part which was thus
visible to me extended, according to the computation given
above, over something more than $4° \frac{1}{2}$ of a great circle. But
I am convinced that, under favourable circumstances of the
weather, situation of Venus, and perfection of instruments, the
atmosphere of Venus might possibly be traced something fur-

ther : this, however, has not been done, at least as yet ; for if we compare with these measurements and calculations, which are certainly as accurate as I could make them,

(γ) *Dr.* HERSCHEL's *observation of the 20th May,* 1793, when he measured the projection of the horns beyond a semicircle, in the evening likewise, about half past eight, *but a day earlier* than I did ; it will be seen that he determines the magnitude of this projection on a mean from two measurements, *with the extreme exactness of* DECIMAL PARTS *of a second,* to be 18° 9′ 8″,2. But this mean is from two measurements which differ from each other, not barely by seconds or minutes, but by MANY DEGREES. In order to judge of the dependance to be placed on them, I will consider each of his measurements separately.

$$
\begin{aligned}
\text{Ist measure. log. } 500 &\quad 2,6989700 \\
\text{log. } 1195 &\quad 3,0773679 \\
\hline
&\quad 9,6216021 \\
=\ &24°\ 44′\ 3″
\end{aligned}
$$

$$
\begin{aligned}
\text{IId measure. log. } 620 &\quad 2,7923917 \\
\text{log. } 2400 &\quad 3,3802112 \\
\hline
&\quad 9,4121805 \\
=\ &14°\ 58′\ 18″
\end{aligned}
$$

His two measurements, therefore, give separately, the first 24° 44′, the second only 14° 58′. *An enormous difference of almost ten degrees,* which, according to my humble judgment, leaves the mean uncertain, not to seconds and their *decimal parts,* nor even to minutes, but properly to 5 degrees. It would therefore be useless to compare further with mine two examples.

which are so little exact, and agree so ill together; and
I must leave it to be judged by others with what reason any
person, from such inaccurate measurements, could consider
mine as erroneous (which besides were made under other circumstances, in the year 1793), and the calculations founded
on them as extremely inexact. Nevertheless, the mean deduced from those examples, namely, 18° 9', agrees very well
with my observation; for the following day, when the projection ought to be greater, I found it 18° 28'; though when it
is considered that the penumbra must be deducted from the
measurement of my opponent, the mean is somewhat too
small. *His observation, therefore, by no means gives the extent of
Venus's twilight greater than mine, but rather something less.*

Thus, by these new measurements and computations, the
general results I have already deduced in my abovementioned
paper " on the atmospheres of Venus and the moon," relative
to the atmosphere of Venus, are still more confirmed and jus-
tified; and there is no longer any doubt, as my opponent
agreeing with me allows, *that the atmosphere of this planet is
very dense, like that of the earth.* Here then I might rest with
regard to those conclusions; however, I find it useful to add
the following explanations, in order to avoid further misunderstanding.

1. Although, according to those results, there is no doubt,
that the atmosphere of Venus is as dense as that of our earth,
yet I do not see in fact, from my observations, how we can con-
found, against all analogy, a general density, with particular,
local and accidental, temporary modifications and condensa-
tions into clouds; and *so positively deny all transparency* to this
atmosphere, as to assert *that in the shining of the planet we see by*

*no means the light of its body, but merely that of its atmo-sphere.*

Notwithstanding the density of this atmosphere, we muşt na-turally consider it as generally clear and transparent, like our own, and that of the moon, and as losing its transparency only where its matter becomes really condensed ; which condensa-tions, however, may be supposed not always to appear like dark-er spots to an observer on our earth, but to remain often imper-ceptible to him. At least, I cannot think, contrary to all analogy, that Providence would bless the inhabitants of Venus, incom-parably less than ourselves, with the happiness of seeing the works of almighty power, and of discovering, like a HERSCHEL, still more and more distant regions of the universe. We must, at least, adhere to this analogy, till indisputable experiments convince us of the contrary, which, however, according to my numerous observations, is by no means the case.

2. But if the atmosphere of Venus be naturally clear and transparent, like that of our earth, except accidental conden-sations, we cannot well doubt, that in looking at the planet, we perceive at the same time both the light of its body, and that of its atmosphere, the latter being illuminated partly by the immediate rays of the sun, and partly by reflection from the body of the planet, and by refraction.

3. It is also equally reasonable to suppose, that, as we are ourselves enveloped in a thick atmosphere, and must look, from a great distance, through a dense illuminated atmosphere, not only our own atmosphere, but likewise particularly the density of that of Venus, and the light upon it, as also the various re-flections of the light from the body of the planet, and its re-fractions, will put such impediments in the way, and occasion

such indistinctness, *that we never can distinguish, as we do in the moon, a projection of the land on the surface of the planet, nor even the shadows cast by its mountainous inequalities, unless it be under a combination of every favourable circumstance, and even then only in a faint undefined manner.* This will be more readily apprehended, when we consider, that the shadows on Venus must appear, from the density of her atmosphere, and its reflection and refraction of light, only *dark-grey*, like those on the earth, and not *black*, as they are on the moon.

4. Yet, in the same manner as in the moon, we discern in Venus, even under the most favourable circumstances, only those parts of her surface, which lie nearest to the boundary of illumination, at the time when we see her half enlightened, because then we look, in a shorter line, perpendicularly through her atmosphere, and moreover the reflection and refraction are much less injurious, and the shadows are longest. Only at such times, and when the atmosphere is likewise clear over such parts of her surface, can we see these shadows, which do not appear sharply terminated, but like a faint mixture of greyish shade and light, sensible enough, but not clear.

5. Granting this rational theory, so conformable at least to our experience on this earth, and to analogy, all the phæ-nomena I have pointed out are very easily and clearly explained by it ; and this experience shews at the same time the justness of the theory, and that it cannot well be otherwise.

Thus we can naturally account for,

(*a*) The soft mixture of light and shade, to be seen only near the time of the greatest elongations, *yet not always, but only sometimes,* and at those moments alone when the atmosphere there, and our own, are favourable for the purpose : to this belong also the shadows sometimes seen by me at the southern horn, and which separated the extreme point of it wholly, or in part. It is possible likewise, that the atmosphere may be clear in one place alone of the boundary of light, in which case we should see something of a shadow there only, the boundary line appearing in the other parts as usual, not streaked with shade, but only not sharply terminated : so, for instance, it was on the 31st of July last year, when Dr. OLBERS observed here with me.

(*b*) But if Venus be considerably more or less than half enlightened, the shadows are not only shorter in themselves, and less perceptible in so small an image, but likewise we see them obliquely, and in a sensibly longer line through the illuminated atmosphere of the planet, which then covering the shadows more, renders them more difficult to be distinguished, and commonly quite invisible. It is, therefore, useless to expect such appearances of shadow, in small crescent phases of Venus, although she be then vastly nearer, and her apparent diameter much larger. If there are at those times real shadows on her, we see the places, not as spots of shade, but as indentations ; and to this belongs the remarkable observation, when the boundary arch of light appears irregular, sometimes in larger and sometimes in smaller parts, and the point of one horn, nay even a considerable part of the horn, is evidently slenderer than the other.

Here it will be readily understood,

(*c*) That as our own atmosphere has an influence on the distinctness of all such phænomena, so accidental condensations in the atmosphere of Venus may cause many bright parts, not lying in the shade, to assume the appearance of dark spots. This accident, however, of which indeed I have no sufficiently certain experience, must occur but seldom, because I have hitherto perceived the mixture of shade and indentation only at the boundary of light; and it would not be easily explained, *why those dark places should not be perceived further in upon the enlightened parts, unless they were true shadows of mountains, and not barely atmospherical appearances.*

Thus at least is every thing to be explained very naturally; and if the phænomena themselves are put out of all doubt by me and others, they confirm the propositions delivered above. And equally insignificant appears to me also, the doubt which

6. A phænomenon might raise, that occurred to my opponent only or chiefly in April of last year : the same, as may easily be supposed, was seen by me many years ago, but especially in 1790, and frequently since; though, not thinking it particularly instructive or remarkable, I forgot to deliver it separately in my paper " on the atmospheres of Venus and the moon."

The phænomenon in question, according to my older observations, consists in this ; that the external edge, for a very small breadth, appears incomparably brighter than the rest of the enlightened part, nearer to the boundary of light; and forms a much brighter small border, which is sharply terminated at its outer edge, but on its inner side appears without any sharp boundary, losing itself in the weak light of the rest of the illuminated part; so that in general, the falling off, or gradual diminution, of the light toward the line bounding the illumination, *is perceived according to the photometrical laws, but particularly becomes chiefly striking nearer to the boundary of light.*

What seems to deserve further attention in this phænomenon, is the circumstance, *that I have seen this extremely brighter
border at the edge, not only about the time of Venus's greatest di-
gressions from the sun, when she appears to us half enlightened, or
more, but also equally well very near the conjunction; and particularly plain in the year* 1790, *when she had the very smallest crescent phase, not amounting to more than from* 4 *to* 6 *seconds in
breadth.*

Were it not for this remarkable circumstance, I should look
for the cause solely in the greater quantity of light, which,
when the planet has the phase of being half, or almost half
illuminated, falls quite or nearly perpendicular through its
atmosphere, on the surface which appears to us the edge, and
is reflected back from this surface into the atmosphere, by
which it is again reflected, and in various ways refracted, so
that at the edge, against which we look by an oblique long
line through the atmosphere, we see an exceeding quantity of
light, *being that of the planet and its atmosphere at the same time;*
but the abovementioned observation seemed to make it probable, that, as I have always believed, the appearance chiefly
depends on optical fallacy, yet this still requires further investigation. However, though we are as little acquainted with
the natural constitution of the ball of the planet, in respect to
its power of reflecting more or less light, as with the species of
the refraction there, yet it seems contrary to all analogy, that
the atmosphere of this heavenly body should be an *opaque*
cover, capable of reflecting more light than the solid body itself; yet that we *should see the external edge, not faintly
expressed, in the manner of an atmosphere, but sharply terminated; and, on the other hand, the boundary of light, under*

*favourable circumstances, streaked with shade,* exhibiting an irregular arch of termination, with indented spots, unequal horns, and so forth. I shall, therefore, at least till adequate reasons convince me otherwise, never assent to a bare hypothesis, that in this planet we merely see its atmospherical cover, and never the body itself; unless when, very rarely, a clearing up of its atmosphere allows us to get sight of a small part of its real surface, in the dark form of a cloud-like spot.

Finally, as to what the celebrated author has remarked besides, *on the apparent diameter of Venus, in the mean distance of the earth;* namely, that by a mean of the measurements he made Nov. 24th, 1791, with the 20-feet reflector, it amounts with great certainty, to 18″,79; and that therefore the planet is larger than it has been given by astronomers hitherto: this is a matter which belongs only indirectly to my object here.

I could have wished that he had not depended too much on a single instrument, having an excess of light, in which the irradiation may unobservedly extend further than in weaker telescopes, nor on a single micrometer; but had reduced his mean from many measurements, made with various and less powerful telescopes, and on many days, under very different apparent diameters, in order to his conclusion for the mean distance of the earth; because, as I only observe here previously, for want of room, I doubt very much of the dependence to be placed on those measures; and must consider this, at least, as rather too large, until I can convince myself of the contrary.

Comparing this determination with that which has been

adopted hitherto, according to M. DE LA LANDE, namely 16″,7, it follows by calculation, that on the 12th March, 1790, when I found the apparent diameter 59 to 60 seconds, it should have been, by M. DE LA LANDE, 58″,58, but by this new determination, 65″,91 ; and on the 21st May, 1793, when I found it greater in proportion, probably because the planet was lower, and had therefore more irradiation, namely 60 seconds, it should by M. DE LA LANDE have been only 56″,75, but by the new determination 63″,85; consequently, according to the latter, I must have overlooked 4 seconds on the 21st May, 1793, and on the 12th March, 1790, when Venus appeared to my eye particularly distinct, *fully 6 seconds.* Both, and especially the last, seem to me contrary to all probability.

As the author, since the year 1780, has measured the diameter 7 different days, so have I before me no less *than 24 different* measurements, made since the year 1788 only : in these I took the apparent diameter of Venus, sometimes when she was at a greater, and sometimes at a less distance ; not only repeating the measurement each time, but often 6, 7, or more times, with different telescopes, magnifying powers, and projection micrometers. If, out of so considerable a number of observations, the mean of the measurements made at each time be taken, and reduced to the mean distance of the earth from the sun, and then the mean of all these reductions be found, this must give the apparent diameter of Venus, at the mean distance, as exactly as possible. Having so great a number of measurements, I must reserve this subject for a particular memoir: yet I think it my duty previously to announce, that in so many observations, I have always found her apparent diameter agree, to 1 or 2 seconds, with that given in the Ephemerides for the time; and as these are computed on the determination hitherto adopted, of $16''$,7, we may continue to reckon Venus of about the same size as she has hitherto been estimated.

Lilienthal,
April 1, 1794.

Fig. 1.

West

North

South

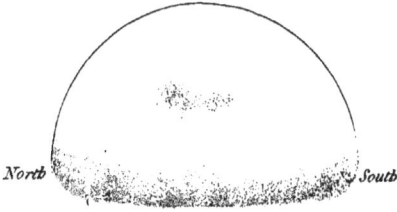

Fig. 2.

a

b

Fig. 3.

a

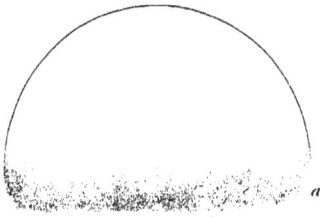

Fig. 4.

b

a

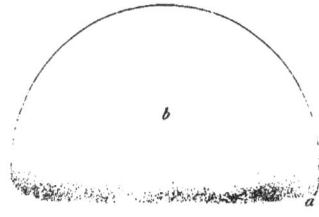

Fig. 5.

a

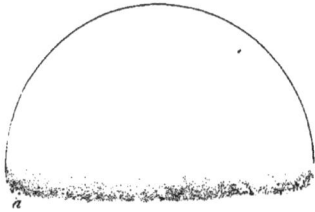

Fig. 6.

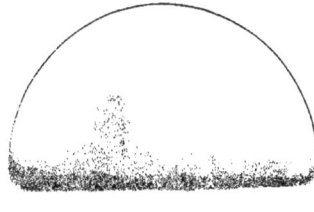

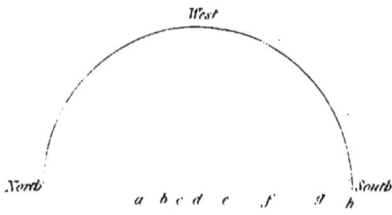

*Fig.7.*

West

North           a b c d e  f  g h  South

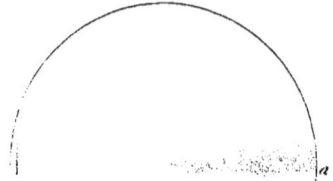

*Fig.8.*

a

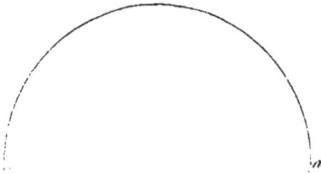

*Fig.9.*

a

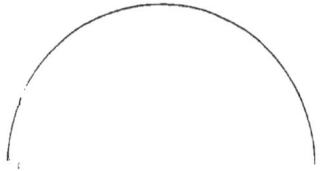

*Fig.10.*

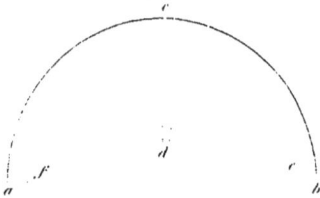

*Fig.11.*

c

f       d      e

a             b

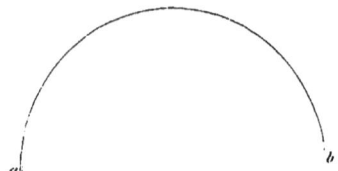

*Fig.12.*

a            b

Fig. 13.

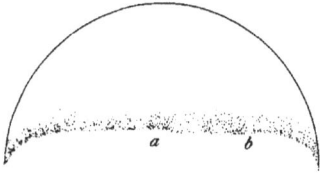

Fig. 14.

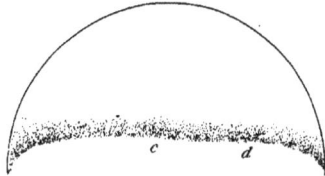

Fig. 15.

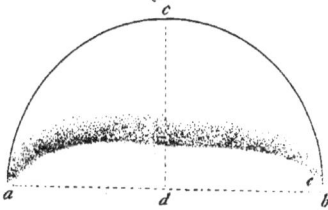

Fig. 16.

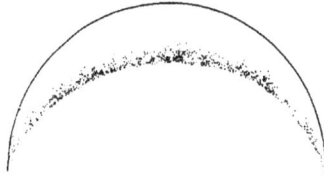

Fig. 17.

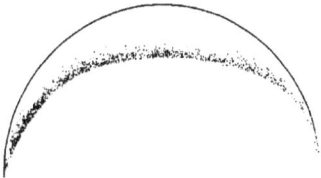

Fig. 18.

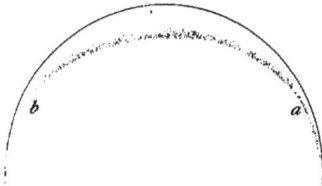

## Fig. 19.

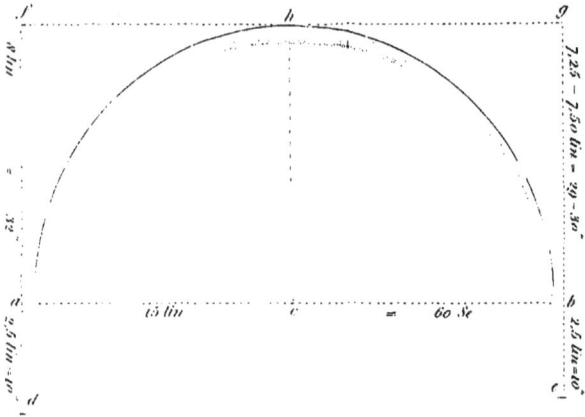

## Fig. 20.

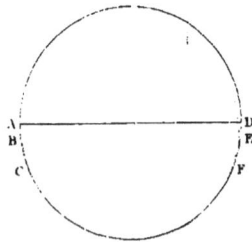

.